Khodjibergenov Dauletbek

Rotary processing of products

Khodjibergenov Dauletbek

Rotary processing of products

ScienciaScripts

Imprint

Any brand names and product names mentioned in this book are subject to trademark, brand or patent protection and are trademarks or registered trademarks of their respective holders. The use of brand names, product names, common names, trade names, product descriptions etc. even without a particular marking in this work is in no way to be construed to mean that such names may be regarded as unrestricted in respect of trademark and brand protection legislation and could thus be used by anyone.

Cover image: www.ingimage.com

This book is a translation from the original published under ISBN 978-3-659-79319-6.

Publisher:
Sciencia Scripts
is a trademark of
Dodo Books Indian Ocean Ltd. and OmniScriptum S.R.L publishing group

120 High Road, East Finchley, London, N2 9ED, United Kingdom
Str. Armeneasca 28/1, office 1, Chisinau MD-2012, Republic of Moldova, Europe
Printed at: see last page
ISBN: 978-620-8-31352-4

Copyright © Khodjibergenov Dauletbek
Copyright © 2024 Dodo Books Indian Ocean Ltd. and OmniScriptum S.R.L publishing group

TABLE OF CONTENTS:

CHAPTER 1

GENERAL CHARACTERISATION OF WORK

Relevance of the work. One of the main tasks of the industry is the wide development of techniques and technology that ensure high productivity and quality of products.

This, first of all, refers to the manufacture of parts and products for various purposes, including those made of hard-to-machine steels and alloys, characterised by high labour intensity and cost of both basic machining operations (turning, milling, and d p.) and finishing and finishing operations (deburring, polishing, etc.). Therefore, in various branches of industry, extensive research is conducted to improve existing machining methods, new high-performance technological processes of forming and finishing of parts are sought and developed.

The tool, penetrating into the workpiece to a certain depth, performs a large amount of deformation work, as the directions of cutting speeds and chip movement have a large divergence.

The front surface of the cutting tool, in the process of separating the stock from the machined surface, redirects it in the opposite direction in the form of a chip. Thus, the separated chip is subjected to large deformation, which leads to increase in cutting temperature, tool wear, etc.

While maintaining the tool strength, we have no possibility to increase the forward angle γ in order to improve the cutting process. In all conventional blade machining schemes, we can identify a cutting wedge characterised by cutting angles and the angle of the wedge itself β.

When selecting the values of these angles, it is necessary to take into account the double requirement for the cutting part design. Thus, ensuring maximum strength of the cutting wedge requires increasing the wedge angle β and selecting the minimum required value of the back angle a.

At present, non-traditional methods of machining, which include rotary machining, are widely used at leading machine-building production facilities. To ensure the required quality of the machined surface, multi-blade rotary machining is proposed
machining performed on lathes and milling machines when machining workpieces made of ductile and hardening materials with deformation prone to outgrowth, as well as when machining complex alloys, heat-resistant alloys, titanium and its alloys.

The method is based on rolling friction between the rear surface of the cutting wedge and the machined surface and does not require significant forces for

2

interaction, which is always permissible. In this case, the tool self-rotation occurs mainly by means of sliding friction between the front and surface of the cutting wedge and the escaping chip. In finishing operations, friction on the rear surface of the cutting tool prevails.

The method allows blade machining to achieve the required quality of the machined surface, while eliminating the need for abrasive machining. This will get rid of charging the machined with small particles of abrasive tool, which subsequently embed in the surface, which adversely affects the wear resistance of the machined part during its operation in machine assemblies. This point is especially important for moving joints.

Output parameters characterising rotary machining during turning on lathes are as follows: roughness of the machined surface $R_a < 0.6$ μm in one pass (roughness value does not matter, machining from "black" surface is possible); relative length of the surface $t_p > 40\%$ at the level starting from the depth of 0.3 mm; durability in comparison with machining with a prismatic cutter increases 5-10 times; temperature in the cutting zone decreases by 30-50%; machining of very hard materials is possible (HRC $\approx$ 60 units.).

For machining unhardened steel and cast iron, as well as for non-ferrous metal it is recommended to use tool material made of high-speed steel (P6M5; P9K8, etc.), for hardened steel - hard alloy BK8 (TK15).

The study of existing problems for conventional milling of ductile and deformation-hardening materials is difficult due to the intermittent nature of the process. It was found that low productivity is observed in deep grooves due to low resistance of discs.
milling cutters. It is difficult to ensure high surface quality. The proposed rotary machining eliminates intermittent cutting, which increases tool durability, and the machined surface is of higher quality.

Among the known metalworking processes, metal cutting is characterised by low energy intensity, which is tens, hundreds of times less than the energy intensity of known physical, chemical and other methods.

Thus, the development of new methods of materials processing and creation of designs of multiblade rotary cutting tools, providing in the process of machining the increase of tool durability, productivity and quality of the machined surface without the use of finishing operations, is relevant.

Relation of the topic to the plan of scientific works. This work is carried out within the framework of scientific and technical programmes on the theme B -

MSF -06-05-06 / 7 "Management of the quality of the machined surface at high-intensity methods of mechanical and physical-mechanical processing" in the state budget research work on the problem of increasing productivity and quality of machining and assembly (cipher 49/7, state registration N 01825049250).

Purpose and objectives of the work research. The aim of the work is to develop economically favourable resource-saving methods of mechanical processing of materials and to create a design of cutting tools, providing an increase in their durability, productivity, improving the quality of the machined surface and reducing energy consumption.

In order to achieve the set goal, the following **tasks** are solved in the paper.

1. To develop scientific bases of the rotational method of processing of products on metal-cutting machines, allowing to calculate the main parameters of the technological system by mathematical models.

2. To offer practical recommendations on the selection of cutting modes and geometric, setting parameters of the tool, which give an opportunity to control the quality of the machined surface on the developed cutting schemes and the method of machining materials.

3. To develop tooling designs and rational design geometric parameters of cutting tools, as well as to create a number of rotary cutting tools for industries.

4. Experimentally prove the multiple increase of tool durability when working according to the proposed schemes and methods of machining.

5. To investigate the shape and dimensions of the chip, temperature and dynamics of the cutting process, tool wear, as well as to develop a methodology and establish dependencies for the calculation of geometric, setting and kinematic parameters of cutting tools and create a new mechanism of chip formation when combining the direction of speed, feed and cutting tool movement, with the development of practical recommendations.

The main provisions of the work put forward for defence:

1. New rotary processing technology.

2. Identified values of quality parameters and cutting modes at combination of hardening and smoothing operations, allowing to increase efficiency of rotary machining.

3. Original methodology of research of geometrical, setting and kinematic parameters of cutting tools.

4. Methodology of calculation of thermal phenomena during rotary processing.

5. Results of experimental studies of temperature, components of cutting forces, tool wear, chip formation process and surface roughness at multi-blade rotary machining.

6. Mathematical dependencies determining the conditions of contact between the cutting part of the tool and the workpiece.

7. Recommendations on selection of cutting modes and geometric and setting parameters of tools, as well as tool materials for rotary tools and tooling.

Scientific novelty of the work consists in the creation of a new scientific direction in the technology of rotary processing, including:

- development of scientific bases of technology of multiblade rotary machining of cylindrical and flat surfaces;
- development of a new method of rotary machining, promising designs of cutting tools and tooling, confirmed by 3 patents of the Republic of Kazakhstan;
- determination of characteristic values for the most rational geometry and kinematics of the cutting process in the proposed method of rotary machining;
- establishment and disclosure of the physical mechanism of the phenomena occurring at the elementary section of the contact zone;
- determination of the relationship between the characteristics of the new method of rotary machining: chip contact length with the front surface, contact stress and contact friction, front and rear tool angle in the cutting process;
- combination of hardening and smoothing operations in the proposed method of rotary machining;
- mathematical modelling of peculiarities of interaction between mechanics of rotary cutting tool and processed material.

The practical significance of the obtained results lies in:

- development and manufacture of new designs of rotary tools and tooling for machining products on metal-cutting machines.
- development of methodical recommendations on selection of cutting modes and geometrical and setting parameters of tools, as well as tool materials for rotary tools and tooling.
- increase of tool life by two orders of magnitude, i.e. increase of durability of multi-blade rotary tool from P6M5 at machining of hardened steels up to $T = 30$ hours;
- $4 \div 8$ times more productivity;
- possibility of controlling the quality of the machined surface with achievement of roughness $R_a < 0,32$ μm at relative length of the supporting surface

t_p = 65 % at the level of the profile cross-section 0,6 mm with the degree of deformation within 0,36 ÷ 0,42 and hardening of the surface layer at the depth up to 0,5 mm;

- reducing the average cutting temperature on the contact surfaces to 200 - 300° C;

- in reducing the consumption of tooling materials;

- in reduction of energy consumption of machining processes and cutting forces (cutting forces are reduced by 15 ÷ 20%);

- exclusion of finishing operations from the technological process of part machining, and obtaining in one pass from the raw surface the accuracy of geometric dimensions within 6...7 qualitets;

- reduction of specific costs of parts manufacturing by more than 2 times;

The results of the work have been approved and implemented in the production of JSC "Kardanval", LLP "Mechanical Plant" (Shymkent) and LLP "Electroapparatus Plant", as well as in the educational process when reading lectures on the discipline "Fundamentals of cutting theory" for students of speciality "Mechanical Engineering" of South Kazakhstan State University named after M. Auezov.

The personal contribution of the co-researcher consists in identifying problems in the field of mechanical processing of materials, in setting tasks and in conducting research, in developing the foundations of the theory of a new method of rotary machining with new cutting tools, in processing and analysing the obtained experimental data. All research work, theoretical and experimental as well as implementation works were carried out personally by the author or with his active participation.

Approbation of the work. The main provisions of the work were reported and discussed at international scientific and technical seminars and national conferences: VII international scientific-practical conference "Naukawa mysl infocrmacyjnej powieki-2011" (Przemysl, Peremyshl, Poland, Nauka I studia 2011g.); VII international scientific-practical conference "Nainovit nauchnochno-postizheniya - 2011" (Sofia, Bulgaria, "Byal GRAD-BG" 2011g.).); International scientific-practical conference "Development of innovative and information technologies in education - the basis of the quality of training specialists", dedicated to the 20th anniversary of independence of the Republic of Kazakhstan, (Shymkent, Republic of Kazakhstan, 2010); International scientific-practical conference "Auezov Readings-9. Ways of innovative development of science, education and culture in the new decade" (M. Auezov SCSU, Shymkent, the Republic of Kazakhstan, 2010); VI International scientific-practical conference "Education and science without borders-2010" (Przemysl Peremyshl, Poland, Nauka I studia 2010); 14th scientific student conference on natural, technical, social-humanities and economic sciences, dedicated to the message of the Head of State "Building the future together" (SCSU, Shymkent, the Republic of Kazakhstan, 2010).

Shymkent, Republic of Kazakhstan, 2011); scientific-practical conference of students, masters, postgraduates and young scientists "New decade - new economic growth - new opportunities of Kazakhstan" (SKSU, Shymkent, Republic of Kazakhstan, 2010); international scientific-practical conference "Perspective directions of alternative energy and energy-saving technologies" (Shymkent, Republic of Kazakhstan, 2010).); International scientific-practical conference "M. Auezov - genius of new time" dedicated to the 110th anniversary of M. Auezov (Shymkent, Republic of Kazakhstan, 2007); Republican scientific-practical conference "The role and tasks of educational institutions in forming the basis of "Smart Economy" (Shymkent, Republic of Kazakhstan, 2007); International scientific-practical conference "Industrial-innovative development - the basis of sustainability of the economy of Kazakhstan" (Shymkent, Republic of Kazakhstan, 2007); International scientific-practical conference "Industrial-innovative development - the basis of sustainability of the economy of Kazakhstan" (Shymkent, Republic of Kazakhstan, 2007).

Shymkent, Republic of Kazakhstan, 2006); international scientific and methodological conference "Improving the relationship between education and science in the XXI century and current problems of quality training of highly qualified specialists" (Shymkent, Republic of Kazakhstan, 2006); international scientific and technical conference "Monitoring of aircraft" (Tashkent, Republic of Uzbekistan, 2000).

Publications on the subject of the dissertation. The results of the dissertation work are reflected in 59 publications, including 31 publications in the editions recommended by the Higher Attestation Commission, there are 3 innovative patents of the Republic of Kazakhstan for the invention of cutting tools.

Structure and scope of the work. The dissertation consists of an introduction, 7 chapters and general conclusions, a list of used sources of 146 names. The material is outlined on 223 pages of typewritten text and supplemented by 5 appendices on 36 pages, contains 86 figures and 3 tables.

CHAPTER 2

MAIN CONTENT OF WORK

The introduction formulates the purpose and objectives of the study, shows the scientific novelty, theoretical and practical significance of the work, presents the provisions put forward for defence. The relevance of the scientific and technical problem is substantiated, the essence is disclosed, the general characteristic and the main directions of the work are outlined. The peculiarities of the processes of mechanical processing of materials are analysed and emphasized. The cutting schemes, which have found the greatest application in the practice of traditional blade machining of materials (Arshinov V.A., Alekseev G.A., Bobrov V.F., Jerusalemsky D.E., etc.) are analysed, which are conditionally divided into the cutting process with inseparable contact between the cutting edge and the processed material and the presence of rotation of the cutting tool having several cutting teeth. On the basis of the analysis of some machining schemes of traditional cutting, advantages and disadvantages in turning, planing, broaching and in milling are formulated. The results are systematised and requirements for the most rational cutting scheme are proposed, which should allow minimising the work of deformation during cutting $A_{деф}$, reduce the friction coefficients ts, reduce the cutting temperature Q, increase the cutting blade front angle, bring the shear angle β_1 closer to 45°.

In traditional cutting schemes, the cutting tool, penetrating into the workpiece to a certain depth, performs a large amount of deformation work, as the direction of cutting and chip speeds have a large mismatch. The front surface of the cutting tool separates the stock layer from the machined surface, stopping it, and directs it in the opposite direction. Thus, the separated chip undergoes a large deformation, which leads to an increase in cutting temperature, tool wear, etc.

The following requirements should be imposed on the cutting pattern: ensuring the possibility to vary the values of cutting tool angles within a wide range while maintaining the wedge angle, keeping it as large as possible; ensuring the continuity of contact

the cutting edge with the material to be machined; renewal of the cutting edge at a speed equal to the chip exit speed. In addition, it is necessary to ensure that the direction of the chip exit velocity and the main motion coincide, and that the allowance to be removed is distributed to the individual cutting edges or elements.

Known methods (Avakov A.A.) of machining work on the principle of sliding between the cutting part of the tool and the escaping chip and machined surface (Novoselov Y.A., Popok A.A.). The speed of relative sliding largely determines both the energy costs of the process (Konovalov E.G., Sidorenko V.A., Souss A.V.) and the

8

tool durability and quality of the machined surface. At the same time, reduction of the relative sliding speed in the contact zones of the tool with the material to be machined can be achieved by replacing sliding at their interaction with rolling.

Cutting schemes that provide, to a greater or lesser extent, the replacement of sliding in the contact zones by rolling are characterised by the highest efficiency. Metal cutting with the replacement of sliding by rolling and the tools that carry out this process are called rotary, because the implementation of rolling in the interaction of the working surfaces of the tool with the material to be machined puts cutting into the class of rotary machining (Konovalov E.G., Sidorenko V.A.).

Thus, the development of scientific bases, technologies and cutting tools for rotary machining of products on metal cutting machines is relevant.

In the first chapter, an analytical review of rotary machining (RM) methods is given, highlighting the main aspects of the cutting process. The existing RO schemes are analysed and research directions are justified.

Some results of researches of E.G.Konovalov, F.V.Bobrov; V.A.Danilov; V.A.Zemlyansky; P.I.Yashcheritsyn and other scientists on the question of difference of mechanics of chip formation, thermophysics, kinematics, dynamics of rotary cutting from traditional methods of mechanical processing are given.

The authors (P.I. Yashcheritzin, A.V. Borisenko, I.G. Drivotin, V.Y. Lebedev) state that due to the long length of the circular cutting edge of the blade, its continuous rotation during operation, as well as good cooling conditions during the idle run, the temperature in the cutting zone during machining with a rotary tool, compared to traditional tools, is reduced by up to 40%. Rotary cutting is also characterised by high durability of the cutting tool, 5-NO times higher compared to traditional blade tools, and the possibility of ensuring high quality of the machined surface.

Konovalov E.G., V.I. Khodyrev, G.F. Shaturov, E.N. Naidyshev prove that relative sliding in contact zones, which is necessary to accompany any known traditional method of machining by cutting, is replaced by rolling with a certain share of slip. Realisation of the said replacement is accompanied by continuous renewal in the process of cutting of the contact surfaces, both of the workpiece and the tool; continuous renewal of the active section of the blade; sharp decrease in the speed of relative sliding in the contact zones. For each elementary section of the cutting blade there is a transition from the process of continuous cutting to periodically recurring, intermittent cutting.

Periodisation of the cutting process improves the working conditions of the tool contact surfaces, it promotes the increase of heat removal from the cutting zone through the tool, the reduction of the overall thermal stress of the process and as a result the tool durability period increases.

Methods of rotary machining based on the principle of periodisation of the cutting edge are known, they are classified in the works of E.G. Konovalov, L.A. Gik, V.I. Khodyrov. Bobrov V.F., Konovalov E.G., Zemlyansky V.A., Kuchma L.K. believe that the main disadvantage for self-rotating rotary cup cutter is slippage.

The formation of a deformation zone during cutting and the evidence for the nature of the deformation is highly controversial, which has led to two directions in the approach to this issue. Perhaps the evidence of the single shear plane model prevails over analytical studies on the
models with a developed deformation zone. Researchers such as Timme, Piaspenen, Merchant, Kobayashi and Thomsen propose a model with a single shear plane. Others such as Palmer and Oxley, Okushimi and Hitomi base their analyses on a model with a developed deformation zone.

It is generally recognised in the theory of machining that sliding friction plays a dominant role in the process of cutting tool wear. It is on this basis that the modern theory of tool wear and durability is based (Konovalov E.G.). It is obvious that in order to increase its wear resistance it is necessary to create conditions under which kinematic friction in the contact zones will be minimised.

The solution of this issue led to the development of a tool with a travelling cutting edge with rear and front kinematic angle equal to zero, since the proposed rotary machining schemes are not considered in the mentioned classification. In addition, the conducted research is valid only for the considered machining conditions and cannot be extended to all RO schemes, RO methods and cutting conditions, the authors remind.

In the second chapter the theoretical substantiation of the scheme of interaction between the cutting tool and the workpiece is carried out. In the same chapter the research tasks are presented.

Creation of resource-saving cutting tools leads to the necessity of specifying some characteristics of the mechanics of interaction of contacting bodies and structural and parametric analysis of the system. On the basis of solving the problem of G. Hertz elasticity theory about compression of perfectly smooth bodies with initial contact along the line and at the point, the conditions influencing the value of the contact area of the tool and the workpiece are considered. The Laplace operator and the Lamé coefficients were used in the modelling of the process under the condition that there is no rest friction.

It is suggested that the contact area can be calculated by the formula

$$S_{\kappa}=P/\sigma_{\scriptscriptstyle T}, \tag{1}$$

where P is the cutting force; $\sigma_{\scriptscriptstyle T}$ is the yield strength of the material being machined.

However, these results corresponded to smooth surfaces. Further contact problems are developed in the direction of modelling the roughness of one or both contacting surfaces on the basis of the Archard rough surface model, taking into account that an increase in cutting force at elastic contact leads to an increase in the contact size.

The condition of transition from elastic contact to elastic-plastic contact is written as follows:

$$h/r \geq 16(\tau_y/E)^2, \tag{2}$$

where h is the layer to be cut; r is the radius of the cutting blade; τ_y is the ultimate tangential stress; E is the modulus of elasticity of the material.

The criterion for transition to plastic deformation is sometimes derived from the condition of existence of some critical value of mean normal stresses at the contact, at which the material completely passes into the state of plastic flow. For a spherical model, such considerations lead to the dependence:

$$a/r = k_n \sigma_\tau (1-\mu^2)/E, \tag{3}$$

where a is the diameter of the contact area; k_n is a coefficient depending on the relationship between σ_τ and the critical stresses on the contact area; μ is the Poisson's ratio.

If the determination of geometrical characteristics of the contact area is considered on the basis of rough surface models, the following basic provisions should be adhered to: contact with rough surface has a discrete character; elementary contacts arise as a result of both elastic and plastic deformations; the contact area and the acting cutting force are related by the following relation

$$S_к = \text{const } N^m,$$

where n - depends on the structure of the model; N - acting load.

The relationship between the mean stresses at the contact of two ellipses and a sphere with equivalent radii is also considered:

$$p_{эл}/p_{сф} = 0{,}7K_1\,[(n+1)n^{-1/2}]^{1/2}, \quad \text{при} \quad n = r_{прод}/r_{поп}, \tag{4}$$

where $p_{эл}$ and p_c f - ellipse and sphere stresses; Ki - coefficient depending on p; Gп po e And Gpop - radius of curvature of the top of the bump in longitudinal and transverse directions, respectively.

The study of contact surfaces of the machined material with the cutter surface is limited to the analysis of contact conditions of smooth and rough bodies. Roughness is modelled by spherical protrusions. In the unloaded state, the rough surface touches the smooth surface with its most protruding roughness. After the application of a load P, the contacting bodies will approach each other by a certain amount AH, and other segments of the rough surface will also come into contact.

Let us select a layer of thickness dx at a distance x from the top of the first of the irregularities that came into contact. All irregularities lying in this layer will approach the smooth surface by the value equal to $(\varepsilon - x)$, where ε is the relative approach of the bodies equal to h_{max}/R_{max}, a $x = X/R_{max}$ - dimensionless coordinate. After some combinations, we obtain an expression for the connection of the acting load

$$P = n_c \int_0^{\varepsilon} P(\varepsilon - x)\varphi'(x)dx , \qquad (5)$$

where R_{max} - cutting tool radius; n_c - insertion depth; $\varphi(x)$ - angle characterising the tool position.

If the function $P(\varepsilon - x)$ is expressed in terms of the mean normal pressure at the contact $p_r(\varepsilon - x)$, and the projection of the area of a unit contact on the plane parallel to the smooth surface is expressed as αSo, then we obtain

$$P = \pi R_{max}\, n_c \alpha\, [2r\int_0^{\varepsilon} p_r(\varepsilon - x)\varphi'(x)dx - R_{max}\int_0^{\varepsilon} p_r(\varepsilon - x)^2\varphi'(x)dx], \qquad (6)$$

where So is the contact area; α is a coefficient depending on the nature of deformation in the contact.

In the obtained relation the values α, $(\varepsilon - x)$, p_r depend on the type of deformation and therefore, in general, it takes different forms depending on the properties of the contacting bodies. Further, in modelling the process, elastic moduli and Poisson's coefficients are introduced and written through the gamma function taking into account the characteristics of roughness distribution: transverse deformation coefficient v and width of the sheared layer b using contour pressure p_c . An equation for calculating the diameter of the elastic contact area d_r is given.

The above considerations allow us to establish the influence of different contact conditions when the bodies are tangentially stationary. However, in rotary cutting, along with normal contact, as a rule, there is also mutual movement of the cutting tool and the workpiece relative to each other.

For further investigation we assume certain conditions: let the two-term friction law take place on the whole contact area: $\tau_{xy} = \tau_0 - k\sigma_y$. Let us consider the force acting on the surface to be machined as normal P and tangential T. In the coordinate system associated with the solid body, the following boundary conditions will take place:

$\sigma_y = 0, \tau_{xy} = 0 \ (-\infty < x < -a, \ b < x < +\infty)$;

$$v = f(x) + \eta, \tau_{xy} = \tau_0 - k\sigma_y \ (-a < x < b), \qquad (7)$$

where v is the normal component of displacements at the boundary y = 0; f(x) is the shape of the contacting solid surface.

Using the Cauchy integral, a special case of the Hilbert-Rimman problem is defined by means of a system of equations that experience the pressure at the contact site.

The obtained relation indicates that the pressure at the contact area depends only on the piezo-coefficient of the molecular component of the friction force k and, therefore, in problems similar to the considered one, it is reasonable to use Amonton's law of friction, i.e. the pressure at the contact area defined as $\tau_{xy} = -k\sigma_y$, can be used at other states of the solid body surface.

Along with other aspects of heat propagation, determining the thermal conductivity of the contacting materials required the use of analytical studies to determine the penetration of the temperature field perpendicular to the treated surface.

When cutting metals, the resulting heat distributed between the workpiece and the tool is dissipated into the environment.

Let us introduce a dimensionless quantity a, which shows what part of the heat flux is directed to one of the contacting bodies. It can be called the heat flux distribution coefficient. Then we can write

$$\begin{cases} Q_1 = \alpha Q \\ Q_2 = (1-\alpha)\,Q \end{cases} \tag{8}$$

where Q_1 and Q_2 are the amount of heat entering the first and second bodies; Q is the total amount of heat generated in the cutting process.

On the other hand, it is known that

$$\alpha = \frac{\lambda_1}{\lambda_2}\sqrt{\frac{a_2}{a_1}}, \tag{9}$$

where λ_1 and λ_2 are the thermal conductivities of the first and second bodies; a_1 and a_2 are the conduction temperatures of the first and second bodies. Taking into account heat transfer, the distribution of heat fluxes will be as follows:

$$1-\alpha = \frac{\sqrt{\pi\sigma^1}}{\sqrt{\pi\sigma^1} + \sqrt{\rho c v}}, \tag{10}$$

where ρ - density; c - specific heat capacity; v - sliding velocity; σ^1 - heat transfer coefficient.

Next, the temperature field during milling of a material that is not heated to the full depth during one working stroke is considered. A strip heat source with a width of 2h and infinitely extended along the y-axis, which is its axis of symmetry, moves along the surface of a semi-finite body (Fig. 1). The heat flux density over the entire surface of the source is assumed to be constant.

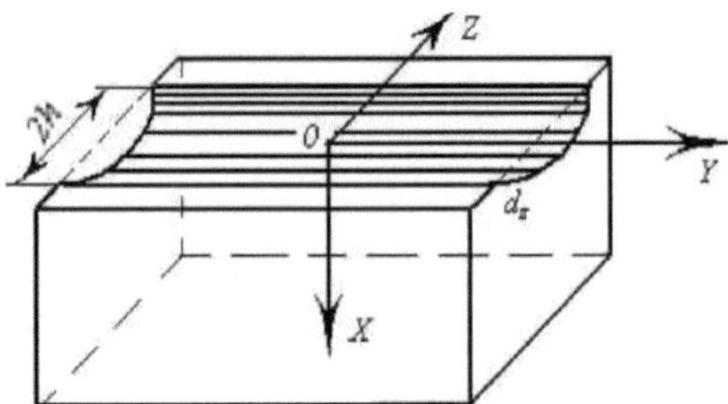

Fig. 1 - Thermal diagram of a strip source

It is reasonable to take the initial temperature of the part equal to zero for simplicity of recording. Let us associate the coordinate system with the heat source. In this case, we can assume that the heat conducting medium moves with the longitudinal velocity in the negative direction of the z-axis. Having adopted such a scheme, we can formulate the problem mathematically. It reduces to the solution of the heat conduction equation or the fundamental solution of the heat conduction equation for a point source on the surface of the part, satisfying the condition of heat transfer of the third kind at the boundary, in a moving coordinate system. In contrast to the previous solution, where a finite amount of heat Q, instantaneously released in an infinitesimal volume, creates an infinitely large temperature at the initial moment, the latter expression describes the process of temperature equalisation from an infinitesimal amount of heat. The temperature at the point of application of such a source is finite even at the initial moment. The whole temperature field will be the sum of such fields from elementary sources distributed over the contact strip.

It is known (Arshinov V.A., Alekseev G.A., Yascheritzin P.I., Feldstein E.) that the main sources of heat during cutting of plastic metals is work. The work is mainly spent on plastic deformations in the sheared layer and in the layers adjacent to the machined surface and cutting surface, as well as on overcoming friction on the front and rear surfaces of the cutter. Applying Lagrange's theorem to the difference of heat inflow using the heat capacity of the cylinder substance c, the density of the cylinder substance ρ ($\rho\Delta xS$ - mass of the cylinder element), equations for determining the heat inflow are given.

Equating the expressions of the same amount of heat $\Delta Q_1 - \Delta Q_2$, we obtain:

$$k\frac{\partial^2 u}{\partial x^2}\Delta xS\Delta t = c\rho\Delta xS\frac{\partial u}{\partial t}\Delta t \quad \text{или} \quad \frac{\partial u}{\partial t} = \frac{k}{c\rho}\frac{\partial^2 u}{\partial x^2}.$$

Denoting $\dfrac{k}{c\rho} = a^2$, we finally obtain the heat propagation equations or the equation of heat conduction in a homogeneous cylinder:

$$\frac{\partial u}{\partial t} = a^2 \frac{\partial^2 u}{\partial x^2}. \qquad\qquad (11)$$

Physical problems are reduced to the problem of heat propagation in a bounded cylinder when the cylinder is so long that the temperature at the internal points of the cylinder at the considered moments of time depends little on the conditions at the ends of the cylinder. If the cylinder coincides with the CE axis, then mathematically the problem is formulated as follows, taking the Poisson integral of heat propagation, transforming it we obtain:

$$u^*(x,t) = \frac{\varphi(\xi)\Delta x}{2a\sqrt{\pi t}} e^{-\frac{(\xi-x)^2}{4a^2 t}}, \quad x_0 < \xi < x_0 + \Delta x. \qquad (12)$$

Formula (7) gives the value of the temperature at a point of the cylinder at any time if at $t = 0$ everywhere in the cylinder the temperature $u^* = 0$ except for the segment $[x_0 , x_0 + \Delta x]$, where it is equal to $\varphi(x)$. The sum of temperatures of the form (6) and gives the solution of the Poisson integral. Note that if ρ is the linear density of the cylinder material, c is the heat capacity of the material, then the amount of heat in the element

$[x_0 , x_0 + Ah]$ at $t = 0$ will be defined as:

$$\Delta Q \approx \varphi(\xi) \cdot \Delta x \cdot \rho \cdot c. \qquad\qquad (13)$$

Let us further consider the function

$$\frac{1}{2a\sqrt{\pi t}} e^{-\frac{(\xi-x)^2}{4a^2 t}} \qquad\qquad (14)$$

By comparing it with the right-hand side of formula (7) taking into account (8), we can obtain the value of temperatures at any point of the cylinder at any time t, if at $t = 0$ there was an instantaneous heat source with quantity $Q = cf$ in section ζ.

In the study of heat propagation in a bounded body it is necessary to add to the heat conduction equation and to the initial condition the conditions on the boundary of the body, which in the simplest cases are boundary conditions of the first, second or third kind. Sturm-Liouville problems with boundary conditions for solving the heat conduction equation are considered. To solve boundary value problems of heat conduction and with inhomogeneous boundary conditions, an arbitrary function subject to determination in the interval of the cutting tool length by the Laplace equation is used. Comparison of the obtained results with experimental data gives grounds to specify the coefficients included in the calculated dependences in order to bring them closer to the real picture realised in the technological process.

The given mathematical calculations allow reasonable planning of experimental studies and interpretation of the obtained empirical results of rotary cutting.

The third chapter discusses the methodology of the research. The solution of problems in the research requires clarification of a number of definitions specific to the rotary cutting tool. The designations of the introduced geometrical and setting parameters of the developed rotary cutting tool are given. These definitions are necessary for the study of a new machining method.

Differences of rotary cutting from traditional methods of machining required the use of analytical and experimental studies, which allowed to determine the actual geometric parameters of the tool and determined the features of the chip formation process, as well as wear of working surfaces and tool durability. In the process of solving the problems, measurements of temperature, force characteristics of the process and study of roughness of the machined surface were carried out.

The experiments were carried out on universal turning and screw-cutting machines of the following models: 1K62, 16K20 (Fig. 2). Special mandrels for rotary turning and tools made of high-speed steels were designed and manufactured (Fig. 3, 4).

Fig. 2 - General view of the experimental setup

Fig. 3 - Mandrel for rotary machining (a) and multi-blade rotary cutting tools made of high-speed steel P6M5 (b), mounted mandrel in the toolholder of a lathe (c), single-blade rotary cutter (d)

The mandrel with a multi-blade rotary cutting tool installed in it is designed for machining the outer surface of bodies of rotation, as well as for face machining. The mandrel is installed in the toolholder of a lathe along with conventional standard cutters (Fig. 3,c).

The mandrel shown in Fig. 3, a mandrel is considered to be a one-position mandrel, because the mandrel holder and the mandrel body are connected by welding. However, other mandrels for changing geometrical and setting parameters were also used during the experiments (Fig. 4).

Fig. 4 - Equipment for carrying out experiments with the possibility of changing the position of cutting elements in the vertical and horizontal plane in the toolholder

To study the nature of chip formation, visual observation was carried out, where the size and shape of the deformation zone can be studied and an external view of how the sheared layer is sequentially transformed into a chip at RO can be made. Initial experiments to study chip formation were carried out at the lowest possible cutting speeds (2...3 m/min) to exclude the influence of other factors of the cutting process (changes in kinematic parameters, temperature, etc.).

In addition, a picture of chip formation can be obtained at low cutting speeds. In this case, the contact magnitude ψ, chip angle and chip length were measured and, by breaking the shear layer to the beginning of the shear line, the shear angle values β_1 . Other parameters of cutting regimes, set angles and tool geometric values were varied so that they did not contribute so much to temperature rise and other factors. Cutting modes, setting and geometric parameters were taken in the following ranges: feed S =

0.07 ÷ 0.95 mm/rev; depth of cut t = 0.25 ÷ 6 mm; tool axis angle in the vertical plane $\beta_y = 5° ÷ 50°$.

The study of chip formation requires the use of workpieces made of ductile and deformation-hardening materials prone to growth, such as aluminium AL70, brass L68 ($\sigma_b = 320$ Pa; NV 55), as well as steel 40X ($\sigma_b = 580$ Pa; NV 183) and steel 45 ($\sigma_b = 610$ Pa; NV 245), grey cast iron SCh32 ($\sigma_b = 520$ Pa; NV 255) widely used in the research. Steel and cast iron were also used for other experiments in turning and facing of external surfaces of bodies of rotation.

In experiments conducted to study the durability of multi-blade rotary cutting tool (MRT) as well as the quality of machined surface, the cutting speed V was varied from 3 to 120 m/min.

The main factor of chip formation - chip shrinkage coefficient was determined according to the known methods. To study the degree of deformation of the sheared layer, the chip shrinkage coefficient cannot fully serve as a quantitative indicator, but gives a qualitative idea. Therefore, experiments were conducted to measure the cross-section of the obtained chips at RO.

Additionally, the pitch of blades co, located parallel in the MRRI, necessary for the removal of cut chips from the cutting zone and for the access of lubricating and cooling process media was determined.

Chip dimensions in cross-section were measured on a BMI-1 tool microscope.

Thermal phenomena in the cutting process are of great practical interest because the heat generated during machining influences the condition of the machined surface, the machining accuracy, and the wear resistance of the cutting tool. The method of natural thermocouple is the most accessible for rotary cutter. In addition, when using it, it is possible to obtain an average temperature in contact of working surfaces with the processed material. The investigations were carried out according to generally known methods (Granovsky G.I., Granovsky V.G., Derganov B.S., Reznikov L.N.) under full electrical isolation.

All main experiments were performed for 3-5 minutes, before blunting of the cutter, without the use of coolant. The thermoelectromotive force was measured for each cutting blade separately on specially made cutting tools.

The constituent forces P_x, P_y, P_z, were measured with a dynamometer UDM-600. Before the beginning of the experiments, the unit was calibrated with a reference dynamometer DOSM-ZM by loading the dynamometer in three mutually perpendicular directions *(x, y, z)* and unloading. The calibration was carried out in different known positions of the TA-5 device.

Sharpening and grinding of the belt on the blade tip were performed on the universal sharpening machine model 312M without cooling. To exclude the total runout of the blade relative to the axis of rotation of the cutting angle, sharpening was carried out in the assembled form, and the mandrel with the grinding wheel was fixed in the three cam chuck of the lathe and dressed with a diamond pencil. The rotary cutting tool was also resharpened. In this case, the total runout of the cutting blade after sharpening of the cutting unit should not exceed 0.01 mm.

In the fourth chapter, the static and kinematic parameters of the rotary cutting tool are considered, as well as the dependences of the thermal and dynamic characteristics of the RO process on the geometry and setting of the cutter.

The process of transforming the cut layer into a chip travelling tangential to the cutting edge is a complicated case of conventional cutting. There are several reasons for this. Prevailing

of which is the rotation of the tool around its own axis. The second reason, which causes the differences, is the kinematic change in the parameters of the cutting part. The third factor that ensures the difference in the chip formation process is the deformation conditions. The latter includes the dimensions of the contact between the tool and the material to be machined.

When in contact with the workpiece, a rotary tool with cylindrical cutting elements creates a curved surface which, by extending, deforms the metal layer to be cut. If the movement of the cutting element is taken into account, the acting forces change in directional magnitude.

For this purpose, the contact of the cutting part of the tool with the workpiece is considered. The point P (y,z) of the cutting element of the rotary tool moves along the line L from the point M to the point N (Fig. 5) The force F is applied to the point P, which changes in magnitude of the direction when the point P moves, i.e. it is some function of the coordinates of the point P: F = F (P).

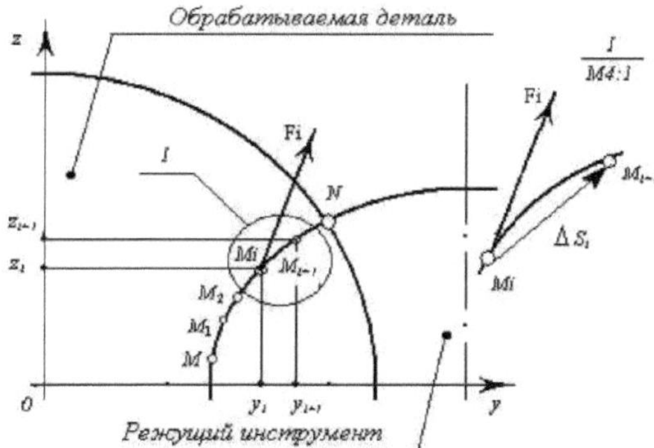

Fig. 5 - Diagram of forces acting on the cutting part of the tool

The work A of the force P is calculated when moving the point P along the contact of the cutting tool (chord MN) with the material to be machined.

19

$$A = \lim_{\substack{\Delta y_i \to 0 \\ \Delta z_i \to 0}} \sum_{i=1}^{n} [Y(y_i, z_i)\Delta y_i + Z(y_i, z_i)\Delta z_i].$$ (15)

In this case, we can apply the curvilinear integral over $Y(y,z)$ and $Z(y,z)$ over the curve L and denote as follows:

$$A = \int_{M}^{(N)} Y(y,z)dy + Z(y,z)dz$$ (16)

The letters M and N, standing instead of limits of integration, are enclosed in brackets as a sign that they are not numbers, but designations of the ends of the line along which the curvilinear integral is taken. The direction along the curve L from point M to point N is the direction of integration.

In our case the curve L is spatial, then the curvilinear integral from the three functions $X(x,y,z)$, $Y(x,y,z)$, $Z(x,y,z)$ is defined similarly:

$$\int_{L} X(x,y,z)dx + Y(x,y,z)dy + Z(x,y,z)dz =$$

$$= \lim_{\substack{\Delta x_k \to 0 \\ \Delta y_k \to 0 \\ \Delta z_k \to 0}} \sum_{k=1}^{n} [X(x_k, y_k, z_k)\Delta x_k + Y(x_k, y_k, z_k)\Delta y_k + + Z(x_k, y_k, z_k)\Delta z_k].$$

The letter L under the sign of the integral indicates that the integration is performed along the curve L. Note that the definition of the curvilinear integral remains valid even when the curve L is closed. In this case, the direction of rotation of the cutting element is necessarily indicated. When the direction of integration changes, the curvilinear integral changes sign, since Δs (Fig. 5) and, consequently, its projections Δx, Δy and Δz change signs.

In the process, the allowance cut by the first blade generates a certain force, which is absorbed by the mandrel spindle axis. Since the back face is involved in cutting, it absorbs certain cutting forces, so there are reaction forces. If the stress $b_K >$ 0, compared to Ru and Pz, the force Px increases dramatically. It can be assumed that of the resulting cutting force Px, the back surface balances a certain part, which must be considered by mathematical methods.

The changes in deformations arising under the action of external forces can be characterised by the displacement vector and, the projections of which on the coordinate axes x, y, z will be denoted as u(x,y,z,t), v(x,y,t), w(x,y,z,t). These displacements arise in an elastic body under the action of internal forces (stresses), which form a symmetric stress tensor.

Considering the volume element and making equations of motion for it, we obtain:

$$\begin{cases} \rho\, \dfrac{\partial^2 u}{\partial t^2} = \dfrac{\partial \sigma_x}{\partial x} + \dfrac{\partial \tau_{xy}}{\partial_y} + \dfrac{\partial \tau_{xz}}{\partial_z} + X\,, \\[2ex] \rho\, \dfrac{\partial^2 v}{\partial t^2} = \dfrac{\partial \tau_{yx}}{\partial_x} + \dfrac{\partial \sigma_y}{\partial_y} + \dfrac{\partial \tau_{yz}}{\partial_z} + Y\,, \\[2ex] \rho\, \dfrac{\partial^2 \omega}{\partial t^2} = \dfrac{\partial \tau_{zx}}{\partial_x} + \dfrac{\partial \tau_{zy}}{\partial_y} + \dfrac{\partial \sigma_z}{\partial_z} + Z \end{cases} \qquad (17)$$

where ρ is the volume density at point (x,y,z); X, Y, Z are the components of external volume forces. The relationship between the stresses arising during deformation and its characteristics is determined by Hooke's law. After some solutions and simplifications we have:

$$\left.\begin{array}{l} P\, \dfrac{\partial^2 u}{\partial t^2} = G\left\{\Delta u + \dfrac{m}{m-2}\dfrac{\partial \theta}{\partial x}\right\} + X\,, \\[2ex] P\, \dfrac{\partial^2 v}{\partial t^2} = G\left\{\Delta v + \dfrac{m}{m-2}\dfrac{\partial \theta}{\partial y}\right\} + Y\,, \\[2ex] P\, \dfrac{\partial^2 \omega}{\partial t^2} = G\left\{\Delta \omega + \dfrac{m}{m-2}\dfrac{\partial \theta}{\partial z}\right\} + Z \end{array}\right\} \qquad (18)$$

To write the previous system of equations as a single vector equation, we use the Lamé equation:

$$P\dfrac{\partial^2 u}{\partial t^2} = (\lambda + 2\mu)\ \text{grad div } u - \mu\ \text{rot } u + F. \qquad (19)$$

We introduce new notations $\mu = G, \quad \lambda = \dfrac{2}{m-2}G$ where μ and λ are Lame constants,

$$t-2$$

because G and m are constants in (17). An arbitrary vector F can always be represented as a sum of

F = grad U + rot L,

where U is a scalar and L is a vector potential.

Let u = grad F + rot Δ,

Where
$$P\,\dfrac{\partial^2 \Phi}{dt^2} = (\lambda + 2\mu)\Delta\Phi + U, \qquad \rho\dfrac{\partial^2 \Lambda}{\partial t^2} = \mu\Delta\Lambda + L,$$

F - volumetric potential; A - vector potential.

It should be noted that by direct substitution and defined in this way the parameter and does satisfy the system of equations (17), which is plausible in the presence of bulk forces. In the equations, the vector potential A in some cases decomposes into three scalar equations. The reduction of the equations to separate scalar equations cannot be seen through to the end without the application or involvement of boundary conditions, which may link different components and thus present considerable difficulty in fully splitting the equations.

If the volume forces are absent, we obtain the homogeneous equations for the potentials F and A

$$\rho\frac{\partial^2 \Phi}{\partial t^2} = (\lambda+2\mu)\,\Delta\,\Phi, \qquad \rho\frac{\partial^2 A}{\partial t^2} = \mu\,\Delta A.$$

The study of cutting patterns (Figures 6 and 7) showed that the following results were achieved:

a) The deformation work is reduced because the direction of the main motion and the feed motion coincide;

б) the work of sliding friction is reduced, as the direction of chip run-off V_c, the resulting speed of the cutting edge of the tool V_e and the rotation speed of the workpiece V_∂ also coincide in direction. And the possibility to regulate the quality of surface machining by means of setting allows to achieve not only the coincidence of the directions of the two first (Fig. 6), but also their equality in value;

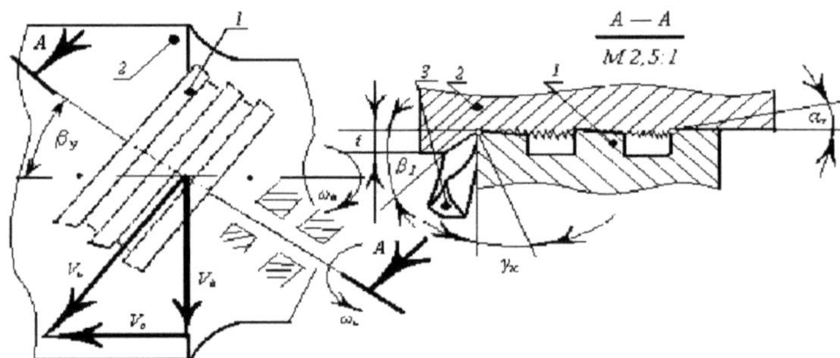

Fig. 6 - Multi-blade rotary cutter for turning bodies of rotation: 1 - cutting elements; 2 - workpiece; 3 - separating chip,
β_y - angle of cutting elements installation; $\beta1$ - shear angle of the cut layer; α_κ - kinematic back angle of the cutting element; γ_κ - kinematic front angle of the cutting element; $\omega_\text{д}$, $\omega_\text{и}$ - angular velocity of the workpiece and cutting elements respectively; $V_\text{д}$, $V_\text{и}$ - linear velocity of the workpiece and cutting elements respectively; V_p - cutting velocity.

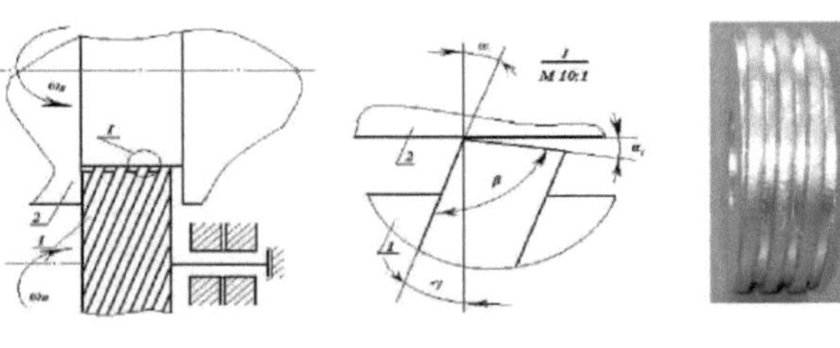

a   *б*

Fig. 7 - Principle of operation of a multi-blade rotary screw cutter: scheme (a) and photo (b) for supporting areas of parts of increased accuracy: 1- cutting elements. 2

- workpiece to be machined. $\omega_\text{д}$ - angular velocity of the workpiece; $\omega_\text{н}$ - angular velocity of the cutting elements; co - helical lift angle; α_c - static back angle of the cutting element; $\gamma_\text{к}$ - static front angle of the cutting element; β - wedge angle

в) sliding friction is replaced by rolling on the rear surface, which is achieved by obtaining a negative rear angle $\alpha_\text{к}$, the correct choice of which will ensure the fulfilment of point "b", and, in addition, the quality of the machined surface can be controlled directly in the machining process;

г) the contact area of the tool with the chip is reduced and the shear (shear angle $\beta 1$) of the chip is facilitated due to the optimum setting angle β_y and the corresponding tool design;

д) it is now possible to change the cutting angles kinematically by means of an appropriate setting (setting angle β_y within a wide range (Fig. 6).

e) progressiveness - the number of cutting edges (cutting elements) is increased with the allocation of roughing, finishing and calibrating sections as required, and) Ensuring high and stable rotor speed equal to the main motion speed. By changing the setting angles β_y and φ of the cutter, the value of the radius of curvature can be adjusted within a wide range.

The cutting edge 4 of the tooth of the developed cylindrical cutter (Fig. 8), in contact with the layer to be cut, easily deforms it in the direction tangential to the cutting feed and takes the cut layer out of the cutting zone in the form of chips. At the same time, the cutting edge 5 of the cylindrical cutter, having the value of the front angle $\gamma = - (22 \div 25°)$, cuts scallops from the machined surface, smoothing its roughness. For supplying the cutting zone with a high-pressure jet of lubricating and cooling fluid, an orifice 6 is provided.

Replacement of sliding friction by rolling friction between the cut layer and the front surface of the cutting edge of the cylindrical milling cutter tooth, as well as the values of the back angle $\alpha = 5 \div 25°$, minimise the build-up and adhesion of the machined material on the cutting edges, excluding sliding between the machined surface and the milling cutter, which leads to an increase in tool durability and productivity of the cutting process.

Continuity of the cutting edge of the tooth in the form of a disc of a cylindrical mill during its movement along the Bernoulli lemniscate trajectory eliminates vibrations, which allows to significantly improve such indicators of surface quality as reduction of deviation from flatness and roughness.

During operation, the static angle α_c of the rear surface of the cutting element in the kinematics $\alpha_\text{к}$ reaches zero.

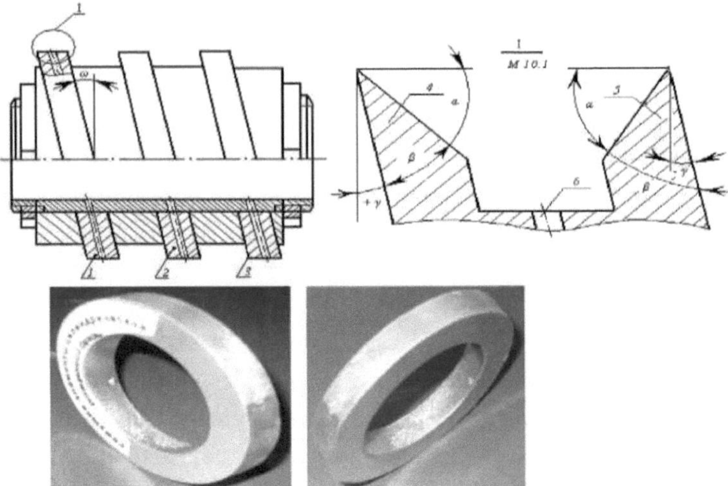

Fig. 8 - Cylindrical rotary milling cutter (a), photo (b) of prototypes of cutting element made of P6M5: ω - angle of inclination of milling tooth; a - static back angle of cutting wedge; y - front angle of cutting wedge; β - wedge angle

Then the kinematic coefficient $K_v = V_и / V_д$ ($V_и$ - tool rotation speed; $V_д$ - workpiece rotation speed) is equal to one, which is proved by experience and analytical means, i.e. the selfrotation process will occur uniformly.

The specificity of the cutting process of the investigated method is such that the shear plane (Fig. 9), ahead of the cutting edge, practically does not contact it. The presence of a real shear plane is explained by the coincidence of the direction of chip run-off $U_{стр}$ with the resultant cutting edge velocity V_e and the replacement of sliding by rolling friction in the contact surfaces. In the process of deformation of the sheared layer, a scheme of plane loading is obtained, which is sometimes commonly referred to as the scheme of "simple shear" (Bobrov F.V.).

From the experiments on measuring the shear angle β_1 it can be seen that at some certain values of the setting angle β_y and feed S the shear angle $\beta1$ reaches approximately 45°.

In case the shear angle is $\beta_1 \approx 45°$, a "pure shear" occurs. The reduction of the intensity of plastic deformation of the cut layer and friction on the front surface of the tool, caused by self-movement of the cutting edge around its axis, should affect the reduction of the cutting temperature.

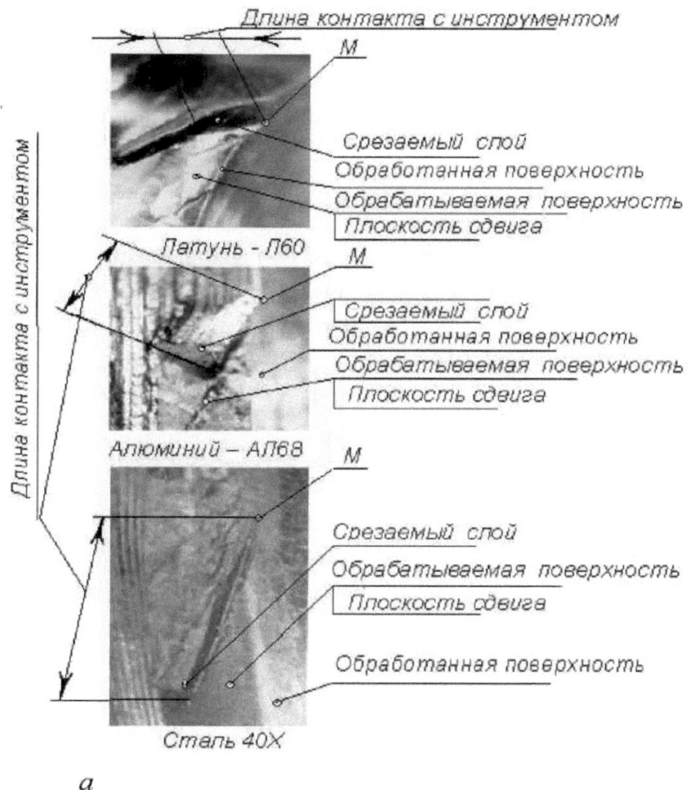

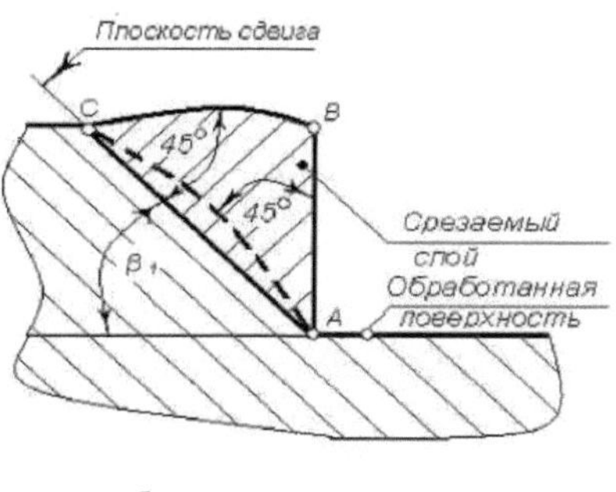

Fig. 9 - The character of the shear plane during material processing (a) and the scheme for determining the chip shear angle β1 (b)

The most typical results of force and temperature measurements are shown in Fig. 10.

The fifth chapter describes deformation processes and chip formation during RO. The theoretical study of the calculation of the active arc of the cutting edge, as well as the determination of the kinematic parameters of the rotary tool are given. Due to the curvilinearity of the cutting blade, it is difficult to determine the dimensions of the contact area (Fig. 11).

All installation parameters, except β_y , are considered in kinematics. After specifying the coincidence of the chip angle and the value of the setting angle $\beta_{yк}$ (Fig. 12, d), it is necessary to determine the values of P_y in using the chip velocity V_{cm} - Sin β_y = KM/KM consequently $KM' = KM \sin \beta$;$_yK'M' = L_a \sin \beta_y + S$;$K'M' = V$ cm-

If a rotary cutter mounted at an angle p_y to the axis of the workpiece is rotated around its axis in addition to the feed motion, the working angle (kinematic angle) $\beta_{yк}$ will be equal to the angle β_{ys} (Bazin V.T.).

$$\beta_{ys} = \beta_y - \operatorname{arctg}\frac{S}{V}, \quad \text{then}$$

$$tg_{yk} = \frac{V_{cm}\cos\beta_y S}{\sqrt{V^2 + S^2 + V_{cm}\sin\beta_{ys}}} \tag{20}$$

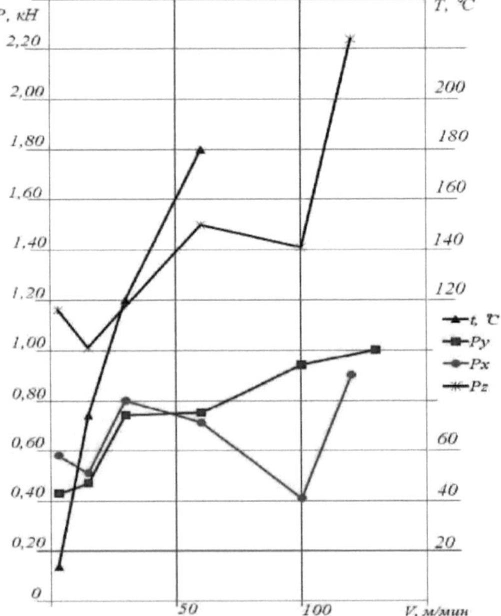

Fig. 10 - Dependence of temperature and components of cutting forces on cutting speed: cutting modes: $\beta_y = 23°$; t = 1mm;$\alpha = 0°$, single-blade RR, point "M" in the centre, machined material steel 45, tool material P6M5.

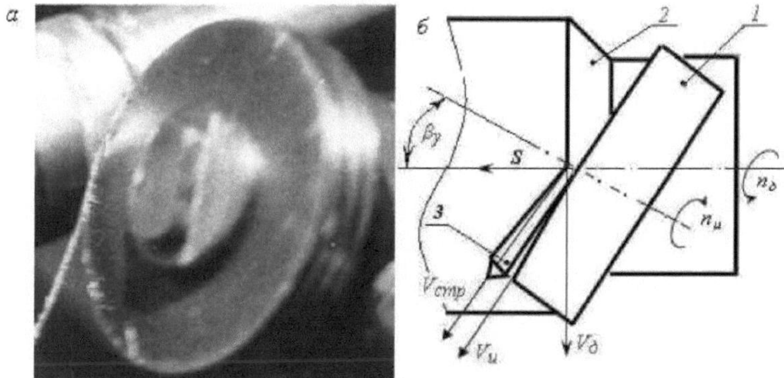

Fig. 11 - Process (a) and scheme (b) of cutting by multi-blade rotary tool: 1 - cutting elements; 2 - workpiece; 3 - chip separating, βy - angle of installation of cutting elements; ω, ω, - angular velocity of workpiece and cutting elements respectively; $V_д$, $V_и$ - linear velocity of workpiece and cutting elements respectively.

The main focus of the chapter is on the issue of chip formation during RO. At the beginning of the contact between the tool and the workpiece at point M (Fig. 9), the tool starts to shear the stock with a small thickness, and at the end of the contact, or at the exit, the sheared layer of metal takes the form of a chip. In RO, due to self-rotation (no slip) of the tool, the chip undergoes deformation on one side only, and the result is a triangular chip shape in cross-section (Fig. 9, b), where the side $\overline{AC}$ is on the opposite side relative to the tool sliding on the chip shear surface. We will call the free side $\overline{CB}$, since it does not come into contact with anything anywhere. The cutter side is the edge of the triangle $\overline{AB}$. In all positions of the setting angle, the surface of $\overline{AB}$, is shiny and even.

The reason for this is that the displaced chip rests AGAINST the front surface of the cutter with the side $\overline{AB}$. The triangle side $\overline{AC}$ is straight in cross-section. The free side of the triangle $\overline{CB}$ is concave.

The area of the triangle is proportional to the feed S, the depth of cut t and the cutting speed V. The chip angle has the same values as the setting angle $β_{ук}$. Since the separated chip is directed towards the direction of the workpiece speed and the front surface of the tool does not allow it to do so, the chip shrinkage increases as the setting angle $β_у$, increases.

The chip shrinkage coefficient KL gives a qualitative indication of the degree of deformation of the sheared layer, but cannot fully serve as a quantitative indicator. Therefore, we have studied the cross-section of the chip, its shape, hardness and the cross-warping coefficient K $_{.п}$

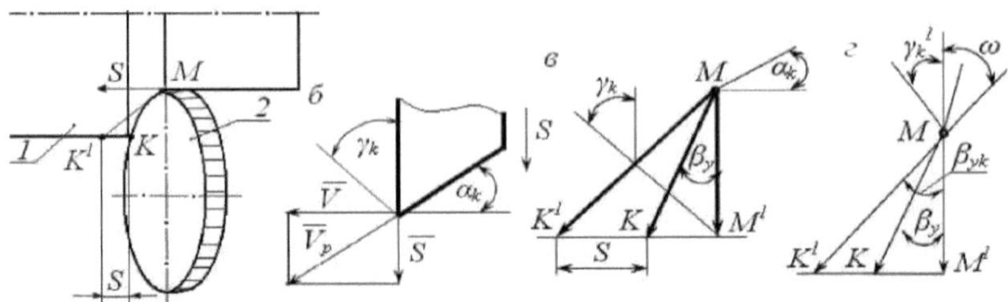

Fig. 12 - Schemes of determination of kinematic parameters of rotary tool: 1-workpiece; 2 rotary tool

The setting angle β_y , influencing the deformation conditions of the sheared layer, obviously influences the size and shape of the chip. The larger the setting angle β_y , the more the chip loses its triangle shape. At installation angles greater than $\beta_y >$ 50° the triangle shape changes dramatically, and the sheared layer of metal turns into a jointed chip, and then into an elemental chip (when machining 40X steel).

At the value of setting angles and cutting modes, in certain ranges, the chip cross-section has the form of a right-angled triangle, and if we take into account the concavity of the side $\overline{AC}$ (Fig. 9), the angle A $\approx$ 45°, and the angle C is also equal to the angle C $\approx$ 45°.

Using the diagram shown in Fig. 9, the shear angle $\beta 1$ can be calculated. The chip side $\overline{CA}$ SLIDES ON THE shear surface and must take the angles $\angle C$ and $\angle A$ on the sliding surface. If $\angle A = 45°$, then $\beta 1$ shear angle is $\beta 1 = 45°$. This confirms that when the shear layer is sheared, the metal failure occurs along the section diagonal, which is equal to 45°. In this case there is a "pure shear" of metal, the least energy is spent, in which case the results of the study of the cross-section of the obtained chips during rotary machining are presented in Figures 13, 14.

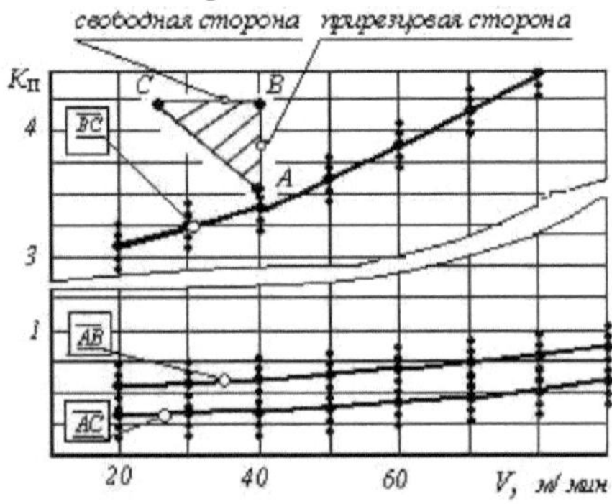

Fig. 13 - Effect of cutting speed V on the coefficient of transverse chip widening $K_п$. at tool setting angle $\beta_y = 20°$; feed S = 0.11 mm/rev; depth of cut t = 1 mm; tool material - P6M5; machined material - steel 40X.

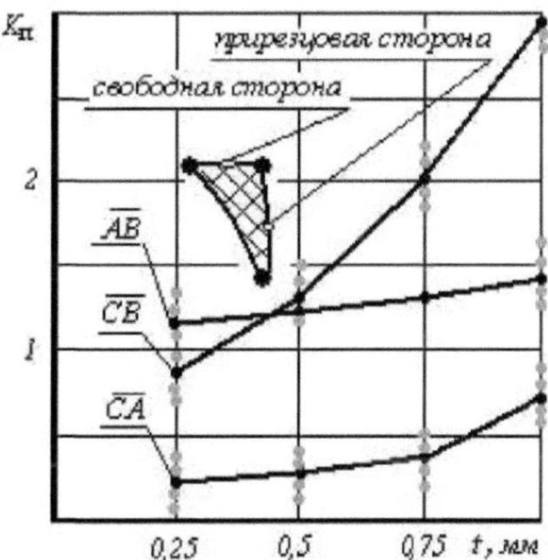

Fig. 14 - Influence of cutting depth t on the coefficient of transverse chip widening $K_\text{п}$: at cutting speed V = 3 m /min; tool setting angle β_y = 30°; feed S = 0,52 mm/rev; tool material - P6M5; machined material - steel 40X.

In any values of cutting speed, the shape of chips is the same, and with increasing speed, the length of LAWS sides monotonically increases. As the cutting depth t increases, the thickness of the cut layer increases. Consequently, the cross-sectional widening of the chip is expected. As S increases, the cross-sections of the obtained chip cross-section ΔABC increase.

The measurement results show that as the installation angle increases, the deformation of the sheared layer increases, since at installation angles β_y = 10 - 20° the hardness does not exceed 260 kG/mm^2 , and at β_y < 25° the hardness is greater than 290 kG/mm .^{2}

The greater the depth of cut, the more difficult it is to shift the layer, as the length of the sides of the triangle $\overline{AC}$ and $\overline{BC}$ (Fig. 9) increases sharply and an obtuse angle triangle is formed.

In the sixth chapter, the issues of durability of cutting tools at RO are highlighted. Experimental studies were carried out to address the above issues and to determine the wear features of the rotary tool.

The length of wear chamfer on the back surface equal to 0.35 - 0.4 mm and the working time - 30 hours at V = 50 m/min; S = 0.1 mm; β_y = 22°; β =20° were recommended as a criterion for blunting of high-speed cutting tools.

Due to the design, several cutting blades can be arranged for roughing, finishing and surface finishing (Fig. 2, 3, 4). The progressive, multi-blade design of these rotary cutting tools makes it possible to increase their durability, compared to a single-blade cutter, because the cutting depth t can be distributed evenly between the blades, on all available blades. Multi-blade allows combining cutting operations and surface-plastic deformation. Practical recommendations for sharpening of cutting tools are given.

The high dimensional stability of rotary tools makes it possible to use them for machining products with high requirements for geometrical accuracy. This machining technology is of particular practical value if the dimensions of the workpiece are significant.

The analysis of the results revealed the peculiarities of the tool wear character in comparison with the traditional cutter and fully confirmed the stated assumptions.

In the seventh chapter, the results of research showing the characteristic features of the effect of rotary machining on the quality of the machined surface are presented. The resistance to deformation changes according to different patterns. With increasing cutting speed, cast iron is de-strengthened without noticeable changes in intensity. Structural steel 45 is also generally subject to de-strengthening with increasing speed due to the increase in temperature Q. For 12X18HIOT steel, the same dependence is observed with a noticeable change in the de-hardening intensity. In the range of $V > 0.5$ m/s, hardening is observed.

CHAPTER 3

GENERAL CONCLUSIONS AND RECOMMENDATIONS

As a result of the carried out researches with the purpose of increase of quality of products the actual problem in the field of development of new methods of processing of details by cutting is solved.

The following conclusions can be drawn from the results of the work.

1. A resource-saving method of machining has been developed for turning by rotary cutting tools and milling, providing high quality of the machined surface, increasing the durability of the cutting wedge, using internal resources of cutting and kinematic parameters of the cutting tool with low energy consumption.

2. Original constructions of multi-blade rotary tools are created and rational conditions of their operation are determined, excluding finishing, finishing and finishing operations, for machining hard-to-machine workpieces from ductile and hardening materials under deformation, prone to outgrowth formation.

3. The peculiarities of mechanics of interaction between rotary cutting tool and processed material are mathematically modelled, structural and parametric analysis of parameters of the created resource-saving equipment for cutting materials is made.

4. Mathematical dependencies for determining:

- work spent on plastic deformations in the sheared layer during rotary processing of materials;

- geometrical characteristics of the contact area of the surface of the machined material with the surface of the cutter, based on modelling representations of a rough surface with moving bodies under pressure;

- regularities of heat propagation at RO of materials for the case of setting the heat flux at the contact boundaries, as well as for arbitrary boundary conditions satisfying the conditions of heat exchange over the whole treated surface, including under the source.

5. Установлено, что разработанная технология ротационной обработки увеличивает срок службы режущего инструмента на два порядка, т.i.e. increases durability of multiblade rotary tool from steel Р6М5 at processing of hardened steels up to T 30 hours, increases productivity of processing (in comparison with traditional) in $4 \div 8$ times; allows to control quality of the processed surface with achievement of roughness $R_a < 0,32$ μm at relative length of the supporting surface $t_p = 65\%$ at the level of the profile cross-section 0,6 mm with the degree of deformation within $0,36 \div 0,42$ and hardening of the surface layer at a depth of up to 0,5 mm; reduces the average cutting temperature on the contact surfaces to $200 \div 300$ °C; reduces the cutting force

(the cutting force is reduced by 15÷ 20%); reduces the consumption of tooling material.

6. The method of calculation of thermal phenomena at multiblade rotary machining is developed.

7. It has been established that sliding friction arising during the machining of products between the machined surface and the tool significantly affects the temperature in the cutting zone, the greater the value of the sliding friction coefficient μ, the more heat is released. Replacement of sliding friction by rolling friction during rotary machining significantly reduces the heat generated and leads to an increase in tool durability.

8. It is proved that reduction of the intensity of plastic deformation of the cut layer and friction on the front surface of the tool caused by self-movement of the cutting edge around its axis ultimately reduces the cutting temperature.

9. It is established that the differences of surface microreliefs (in shape, pitch of microroughnesses and other parameters), processed by grinding wheel and multiblade roundabout tool, are not significant.

10. The results of research are implemented at JSC "Kardanval" and LLP "Mechanical Plant" Shymkent of the Republic of Kazakhstan, as well as in the training centre of the Republic of Kazakhstan
process on the discipline "Fundamentals of cutting theory" for students of speciality "Mechanical Engineering" of South Kazakhstan State University named after M. Auezov.

CHAPTER 4

THE MAIN PROVISIONS OF THE DISSERTATION WORK ARE REFLECTED IN THE FOLLOWING PUBLICATIONS

1. **Khodjibergenov D.T.** Relationship between the front angle y and shear angle P during rotary machining [Text] / D.T. Khodjibergenov, G.S. Mukhamadiev // Izvestiya Vuzov. Journal of the Ministry of Higher and Secondary Specialised Education of the Republic of Uzbekistan. - Tashkent, 2000. - № 4. - C. 3-5.

2. **Khodjibergenov D.T.** Kinematic parameters of the cutting process in rotary machining [Text] / D.T. Khodjibergenov, A. Abdukarimov // Problems of Mechanics. Journal of the Academy of Sciences of the Republic of Uzbekistan. - Tashkent, 2000. - № 1. - C. 66-69.

3. **Hodzhibergenov D.T.** Rotational processing of materials [Text] / D.T. Hodzhibergenov, I.K. Kushnazarov, T.A. Sagdiyev // Aircraft Monitoring. International scientific and technical conference. - Tashkent, 2000. - C. 34-36.

4. **Khodjibergenov D.T.** Quality of a surface layer at cutting and hardening processing [Text] / D.T. Khodjibergenov, I.K. Kushnazarov, M.S. Usmanov // Problems of operation and improvement of transport systems. Interuniversity collection of scientific papers. - SPb, 2001. - C. 111-113.

5. **Hodzhibergenov D.T.** Optimisation of technological operations at performance of aircraft repair works on the basis of ATB in airfield conditions [Text] / D.T. Hodzhibergenov, I.K. Kushnazarov, M.S. Usmanov // Problems of operation and improvement of transport systems. Interuniversity collection of scientific papers. - SPb., 2001. - C. 114-117.

6. **Khodjibergenov, D.T.** Change of contact area parameters at tangential displacement of bodies [Text] / D.T. Khodjibergenov // Aircraft Monitoring. International scientific and technical conference. - Tashkent, 2003. - C. 33-35.

7. **Khodjibergenov D.T.** Thermal problem for the process of multiblade rotary cutting [Text] / D.T. Khodjibergenov // Science and Education of South

Kazakhstan. Republican scientific journal. Series: Processes apparatuses. - Shymkent, 2005. - C. 166-169.

8. **Khodjibergenov, D.T.** Features of mechanics of interaction between the rotary tool and the material of the processed part [Text] / D.T. Khodjibergenov // Mechanics and modelling of technology processes. Scientific and theoretical journal. Taraz, 2005. - C. 224-226.

9. **Khodjibergenov D.T.** Quality management of the machined surface [Text] / R. Alsherova, Sh. Orazymbetov, D.T. Khodjibergenov, A.K. Zhusipbekov //Strategy of Kazakhstan's entry into the 50 most competitive countries of the world. 9th Student Conference on Natural, Social-Humanitarian and Economic Sciences. - Shymkent, 2006. - C. 254-266.

10. **Khodjibergenov D.T.** Management of the machined surface quality at multiblade rotary machining [Text] / D.T. Khodjibergenov, O.B. Kambarova // Search. Kambarova // Poisk. Scientific application of the international journal. "Higher School". Kazakhstan Ministry of Education and Science of the Republic of Kazakhstan. Series of natural and technical sciences. - Almaty, 2006. - C. 274-276.

11. **Khodjibergenov D.T.** Effect of cutting modes on durability and wear during rotary machining [Text] / D.T. Khodjibergenov, S. Abdraimov, O.B. Kambarova // Search. Kambarova // Poisk. Scientific application of the international journal. Series of natural and technical sciences. "Higher School" Kazakhstan Ministry of Education and Science of the Republic of Kazakhstan. - Almaty, 2006. - C. 276-280.

12. **Hodzhibergenov D.T.** Analysis of processes of mechanical processing of materials [Text] / D.T. Hodzhibergenov, S. Abdraimov, A.K. Zhusupbekov // Improving the relationship between education science in the XXI century and current problems of quality training of highly qualified specialists. Proceedings of the international scientific and methodical conference. - Shymkent, 2006. - C. 208-209.

13. **Hodzhibergenov D.T.** Method of processing materials [Text] / D.T. Hodzhibergenov, S. Abdraimov, A.D. Kadyrbekov // Improving the relationship between education science in the XXI century and current problems of quality training of highly qualified specialists. Proceedings of the

international scientific and methodical conference. - Shymkent, 2006. - C. 209-210.

14. **Khodjibergenov D.T.** Research of influence of cutting modes at multiblade rotary machining on the quality of machined surface [Text] / D.T. Khodjibergenov, V.N. Pechersky // Proceedings of the University. Karaganda: Karaganda State Technical University, 2006. - C. 42-43.

15. **Hodjibergenov D.T.** Influence of kinematic schemes in rotary processing [Text] / D.T. Hodjibergenov, A.K. Zhusipbekov, A.D. Kadyrbekov, J.A. Sadykov // Industrial and innovative development - the basis of sustainability of the economy of Kazakhstan. International Scientific and Practical Conference. - Shymkent, 2006. - C. 444-446.

16. **Hodzhibergenov, D.T.** Multiblade rotary processing of materials [Text] / D.T. Hodzhibergenov, A.K. Zhusipbekov, D.S. Myrzaliev, I.K. Kushnazarov // The role and tasks of educational institutions in forming the basis of "Smart Economy". Republican scientific-practical conference. - Shymkent, 2007. Vol. VI. - C. 100- 102.

17. **Khodjibergenov D.T.** Analysis of deformations of the machined surface at rotary machining [Text] / D.T. Khodjibergenov, A.K. Zhusipbekov, V.N. Pechersky // Science and Education of South Kazakhstan. Republican scientific journal. Series: Mechanics and mechanical engineering. -Shymkent, 2007. - № 2 (61). - C. 146-149.

18. **Khodjibergenov D.T.** Analysis of the cutting temperature distribution during rotary machining of materials [Text] / D.T. Khodjibergenov, A.K. Zhusipbekov, V.N. Pechersky // Science and Education of South Kazakhstan. Republican scientific journal. Series: Mechanics and mechanical engineering. - Shymkent, 2007. - № 4 (63). - C. 31-34.

19. **Hodzhibergenov D.T.** Analysis of kinematic parameters of rotary tool [Text] / D.T. Hodzhibergenov, A.K. Zhusipbekov, J.A. Sadykov // M. Auezov - genius of new time. International scientific and practical conference. - Shymkent, 2007. Vol. No. 10. - C. 120-123.

20. **Khodjibergenov D.T.** Metaldardy zertteu zhane sonau tesider! [Text] / A.K. Zhusipbekov, D.T. Hodzhibergenov, J.A. Sadykov, D.S. Myrzaliev // Otsu

kuraly. - Shymkent: M.Euezov atyndagy Oztust!k K,azakstan Memlekettzh University, 2007. -108 б.

21. **Khodjibergenov D.T.** Analysis of mechanical processing processes [Text] / A.Tokhtamuratov, D.T. Khodjibergenov, A.K. Zhusipbekov // Scientific and Technical Space. Proceedings of the scientific-practical conference of students, masters, postgraduates and young scientists, - Shymkent, 2008. - C. 74-75.

22. **Khodjibergenov D.T.** Strengthening of the machined surface during cutting [Text] / D.T. Khodjibergenov, V.N. Pechersky, A.K. Zhusipbekov // Science and Education of South Kazakhstan. Republican scientific journal. Series: Engineering. - Shymkent, 2008. - № 3(68). - C. 129132.

23. **Hodzhibergenov D.T.** Methodology of experimental study of chip formation in metal cutting [Text] / M.D. Akhmetov, I. Bultachiev, A.J. Umirzakov, D.T. Hodzhibergenov, A.K. Zhusipbekov // Integration of science and production - the creativity of young. Scientific and practical conference of students, masters, postgraduates and young scientists. - Shymkent, 2009. - C. 36-39.

24. **Khodjibergenov D.T.** Temperature measurement in the cutting zone of metals [Text] / N.J. Askarov, K.M. Kistaubaeva, A.J. Umirzakov, D.T. Khodjibergenov, A.K. Zhusipbekov // Integration of science and production - the creativity of the young. Scientific and practical conference of students, masters, postgraduates and young scientists. - Shymkent, 2009. - C. 39-42.

25. **Khodjibergenov, D.T.** Analysis of the shear angle Pi at metal machining by cutting [Text] / D.T. Khodjibergenov, V.N. Pechersky, A.K. Zhusipbekov // Mechanics and Modelling of Technology Processes. Scientific-theoretical journal. Taraz, 2008. - № 2. - C. 190-192.

26. **Khodjibergenov D.T.** Heat spreading at multiblade rotary metal working [Text] / D.T. Khodjibergenov // Science and Education of South Kazakhstan. Republican scientific journal. Series: Engineering. - Shymkent, 2009. - № 3(76). - C. 90-93.

27. **Khodjibergenov D.T.** Work spent on plastic deformations in the sheared layer at rotary metal processing [Text] / D.T. Khodjibergenov // Science and

Education of South Kazakhstan. Republican scientific journal. Series: Engineering. - Shymkent, 2009. - № 6 (79). - C. 13 8-140.

28. **Hodzhibergenov D.T.** Tooling tooling for rotational turning [Text] / D.T. Hodzhibergenov, A.K. Zhusipbekov, J.A. Sadykov // Perspective directions of alternative energy and energy-saving technologies. Proceedings of the international scientific-practical conference. - Shymkent, 2010. - C. 174-176.

29. **Khodjibergenov D.T.** Development of a methodology for the study of cutting forces during turning by rotary tool [Text] / N. Askarov, D.T. Khodjibergenov, A.K. Zhusipbekov // New decade - new economic growth - new opportunities of Kazakhstan. Scientific and practical conference of students, masters, masters, postgraduates and young scientists. - Shymkent, 2010. - C. 11-13.

30. **Khodjibergenov D.T.** Research of wear methods of cutting tools and machine part surfaces [Text] / N.G. Shadiev, D.T. Khodjibergenov, Zhusipbekov A.K. // Building the future together. 14th scientific student conference on natural, technical, social-humanities and economic sciences. - Shymkent, 2011. - C. 13-15.

31. **Hodzhibergenov D.T.** Influence of various factors on the forces Px, Ru, Pz during turning with a rotary cutter [Text] / D.T. Hodzhibergenov // Education and Science without Borders-2010. Materials of VI International Scientific and Practical Conference - Przemysl (Poland), 2010. - C. 21-25.

32. **Khodjibergenov, D.T.** Effect of cutting modes on the transverse chip shrinkage at multiblade rotary machining [Text] / D.T. Khodjibergenov // Engineering Technics of mechanical engineering. Scientific and technical journal. -M.: Mashizdat, 2011. № 1 (77). - C. 13-16.

33. **Khodjibergenov D.T.** Cutting schemes of blade machining processes [Text] / B.M. Sunnatov, D.T. Khodjibergenov // Scientific Works of South Kazakhstan State University named after M. Auezov. - Shymkent. - 2010. - № (1) 19. - C. 185-189.

34. **Hodzhibergenov D.T.** Progressive method of obtaining the part [Text] / B.M. Sunnatov, D.T. Hodzhibergenov // Auezov readings - 9: Ways of innovative development of science, education and culture in the new decade.

Proceedings of the international scientific-practical conference, - Shymkent, 2010. - C. 279-281.

35. **Hodzhibergenov D.T.** Rotational method of machining the part [Text] / K.T. Sherov, D.T. Hodzhibergenov // Izvestiya Vuzov. Bishkek, 2010. - № 9. -C. 3-5.

36. **Khodjibergenov D.T.** Temperature measurement in the cutting zone during rotary machining [Text] / B.M. Sunnatov, D.T. Khodjibergenov, N.G. Shadiev, D.Y. Yusvalieva // Development of innovative and information technologies in education - the basis of the quality of specialist training. - Shymkent, 2011. - C. 237-239.

37. **Khodjibergenov D.T.** Analysis of kinematic schemes of materials cutting [Text] / B.M. Sunnatov, D.T. Khodjibergenov, A.K. Zhusipbekov // Renewed Kazakhstan in the world space achievements and prospects of development. Republican scientific-practical conference. - Shymkent, 2011. - C. 306-308.

38. **Hodzhibergenov D.T.** Features of multi-blade rotary cutting tool [Text] / B.M. Sunnatov, D.T. Hodzhibergenov // Naukawa mysl infocrmacyjnej powieki-2011. Proceedings of the VII International Scientific and Practical Conference. - Przemysl (Poland), 2011. 22-27.

39. **Khodjibergenov D.T.** Strengthening treatment of surfaces by plastic deformation [Text] / D.T. Khodjibergenov, K.T. Sherov // Izvestiya Vuzov. Bishkek, 2010. - № 9. - C 9-11.

40. **Khodjibergenov D.T.** Plastic deformations in the sheared layer at rotary metal working [Text] / B.M. Sunnatov, D.T. Khodjibergenov // Najovit nauchni postizheniya-2011. Materials of VII International Scientific and Practical Conference. - Sofia (Bulgaria), 2011. - C 26-29.

41. **Hodzhibergenov D.T.** Features of cutting and strengthening tool [Text] / D.T. Hodzhibergenov // Problems of Mechanics. Journal of the Academy of Sciences of the Republic of Uzbekistan. - Tashkent, 2010. - № 3. - C. 66-69.

42. **Khodjibergenov, D.T.** Analysis of temperature phenomena of rotary processing / D.T. Khodjibergenov, K.T. Sherov. // Izvestiya Vuzov. Bishkek, 2011.- № 6.-P. 9-11.

43. **Khodjibergenov D.T.** Cut layer at rotary machining [Text] / D.T. Khodjibergenov // Problems of Mechanics. Journal of the Academy of Sciences of the Republic of Uzbekistan. - Tashkent, 2010. - № 4. - C. 26-29.

44. **Khodjibergenov D.T.** Rotary cutting tool. Patent RK № 24688 [Text] / D.T. Hodzhibergenov, A.K. Zhusipbekov, B.M. Sunnatov // Published 17.10.2011, Bulletin No. 10.

45. **Khodjibergenov D.T.** Patent RK № 24240. Rotary cutting tool [Text] / D.T. Hodjibergenov, A.K. Zhusipbekov, N.J. Askarov // Published 15.07.2011, Bulletin No. 7.

46. **Khodjibergenov D.T.** Patent RK № 24239. Cylindrical cutter [Text] / D.T. Hodjibergenov, A.K. Zhusipbekov, N.J. Askarov // Published 15.07.2011, Bulletin No. 7.

47. **Hodzhibergenov D.T.** Fundamentals of cutting theory: textbook [Text] / D.T. Hodzhibergenov // - Shymkent, 2008. - 114 c.

48. **Hodzhibergenov D.T.** Components of the cutting force in rotary machining [Text] / D.T. Hodzhibergenov, B.M. Sunnatov // Science and Education of South Kazakhstan. Republican scientific journal. Series: Mechanical Engineering. Shymkent, 2011. - №. 4 (90). - C. 131-133.

49. **Hodzhibergenov D.T.** Analysis of the process of cutting of materials [Text] / D.T. Hodzhibergenov, K.T. Sherov. // Izvestiya Vuzov. Bishkek, 2011. - № 9. - C. 19-22.

50. **Hodzhibergenov D.T.** Character of chip formation at rotary machining [Text] / D.T. Hodzhibergenov, B.M. Sunnatov // Mechanics and Modelling of Technology Processes. Scientific and theoretical journal. - Taraz, 2011. - № 2. - C. 259-262.

51. **Khodjibergenov, D.T.** Influence of various factors on the forces P_x , Ru, Pz at turning by a rotary cutter [Text] / D.T. Khodjibergenov // Technics of Mechanical Engineering. Scientific and Technical Journal. - M.: 2012. - № 1 (81).-C. 12-15.

52. **Hodzhibergenov D.T.** Quality of machined surface at rotary machining [Text] / D.T. Hodzhibergenov, K.T. Sherov // Izvestiya Vuzov. - Bishkek, 2012. - № 6. - C 3-5.

53. **Khodjibergenov D.T.** Investigation of the shear angle at rotary machining [Text] / D.T. Khodjibergenov // Metal processing. Scientific, technical and industrial journal of SB RAS. - Novosibirsk, 2012. - № 5 (55). - C. 22-27.

54. **Khodjibergenov D.T.** Thermal conductivity of the processed material during rotary metal processing [Text] / K.T. Sherov, D.T. Khodjibergenov // Science and new technologies. - Bishkek, 2012. - № 6. - C. 9-11.

55. **Hodzhibergenov, D.T.** Perspective approaches to materials processing by cutting [Text] / D.T. Hodzhibergenov // Mashinostroitel. Industrial scientific and technical journal. - M.: 2012. - № 4. - C. 37-42.

56. **Khodjibergenov D.T.** Resistance of rotary tool [Text] / D.T. Khodjibergenov, K.T. Sherov // Science and new technologies. - Bishkek, 2012.-No. 6.-C. 3-5.

57. **Hodzhibergenov D.T.** Kesu theoryasynyts nepzder! [Text] / D.T. Hodzhibergenov // Okulpu - Shymkent, 2012. - 204 б.

58. **Khodjibergenov D.T.** Kinematics and dynamics of rotary machining process [Text] / D.T. Khodjibergenov // Monograph. - Shymkent: South Kazakhstan State University named after M.Auezov. M. Auezov, 2012. - 172 c.

59. **Khodjibergenov D.T.** Investigation of kinematic coefficient at rotary machining [Text] / K.T. Sherov, D.T. Khodjibergenov et al. // Seitenovskie readings - 7. Materials of the International Scientific and Practical Conference. - Kokshetau, 2012. - C. 163-165.

60. **Khodjibergenov D.T.** Calculation of thermal problem for the process of multi-blade rotary cutting [Text] / K.T. Sherov, D.T. Khodjibergenov // IP RK № 0009167. Record in the register № 1321 from 16.11. 2012.

SUMMARY

Dissertation of Khodjibergenov Davlatbek Turganbekovich on the theme: "Development of scientific bases, technology and cutting tools for rotary

machining of products on metal-cutting machines" for the degree of Doctor of Technical Sciences on speciality 05.02.08 - technology of mechanical engineering

Keywords: cutting, shear angle, degree of deformation, cutting pattern, kinematic coefficient, sheared layer, surface, hardness, roughness, turning, finishing, tool wear, sliding friction, cutting speed, shear plane.

Object of research: technological process of mechanical processing of products by turning and milling.

Purpose of work: Development of economically favourable resource-saving methods of mechanical processing of materials by creating a design of cutting tools, auxiliary mandrels, providing an increase in the durability of cutting tools, productivity, improving the quality of the machined surface and reducing energy consumption.

Research methods and apparatus: scientific and experimental studies were carried out on the basis of the basic provisions of mechanical engineering technology, theory of cutting, mechanics, physics, mathematical statistics, and with the help of standard, information and measuring equipment with the use of computer technology.

The results obtained and their novelty: scientific bases of processing technology have been developed, as well as a new method of multi-blade rotary machining, confirmed by 3 patents of the Republic of Kazakhstan; the physical mechanism of phenomena occurring at the elementary section of the contact zone has been established; the characteristics of problematic types of machining have been determined: the length of chip contact with the front surface of the tool, contact stresses and contact friction, as well as the role of the front and rear angle of the tool in the machining process; the possibility of combining the tool with the front surface of the tool has been substantiated; the possibility of combining the tool with the front surface of the tool has been determined.

The practical significance of the work consists in the fact that a new design of rotary tools and tooling for the processing of products on metal-cutting machines has been developed and produced; methodical recommendations for the selection of cutting modes and geometric, setting parameters of tools, as well as tool materials for rotary tools and tooling have been developed.

Recommendations for use: the developed technology allows to achieve the required quality of the treated surface, while eliminating the need for abrasive treatment.

Field of application: technological processes in metal cutting, the method can be implemented on lathe, planer and milling machines.

yes

I want morebooks!

Buy your books fast and straightforward online - at one of world's fastest growing online book stores! Environmentally sound due to Print-on-Demand technologies.

Buy your books online at
www.morebooks.shop

Kaufen Sie Ihre Bücher schnell und unkompliziert online – auf einer der am schnellsten wachsenden Buchhandelsplattformen weltweit! Dank Print-On-Demand umwelt- und ressourcenschonend produzi ert.

Bücher schneller online kaufen
www.morebooks.shop

Printed by Books on Demand GmbH, Norderstedt / Germany